Anastasia Shchurenko

Building an ecosystem and infrastructure for smart confectionery Book 2

Anastasia Shchurenko

Building an ecosystem and infrastructure for smart confectionery Book 2

Innovative transformation of the ecosystem and infrastructure of a smart composite manufacturing facility

ScienciaScripts

Cover image: www.ingimage.com

This book is a translation from the original published under ISBN 978-620-7-47295-6.

Publisher:
Sciencia Scripts
is a trademark of
Dodo Books Indian Ocean Ltd. and OmniScriptum S.R.L publishing group

120 High Road, East Finchley, London, N2 9ED, United Kingdom
Str. Armeneasca 28/1, office 1, Chisinau MD-2012, Republic of Moldova, Europe
Printed at: see last page
ISBN: 978-620-8-25657-9

Anastasia Shchurenko

Book - 2

Caption

Building a smart confectionery production ecosystem. Book 2

Subheading

Innovative transformation of the ecosystem and infrastructure of a smart facility for the production of composite products

Keywords:

Confectionery production, Digital ecosystem technologies for confectionery production, Industrial design rules and criteria as applied to advanced, smart confectionery production, Application of electromagnetic resonance sensors, System search, Durability of new product.

Annotation

Innovative transformation of the operational and performance characteristics of the ecosystem and infrastructure of a smart facility for confectionery production into an advanced combined supersystem with interconnected on-line monitoring and control subsystems in real time, including energy and water treatment modules.
The importance of applying the principles of innovative design, manufacturing and operation of the whole complex and components of the ecosystem of the infrastructure of the smart object of confectionery production in the formation of a complex system of regeneration and recycling of waste water and other environmental resources with the introduction of innovative systems of technological production relaxation in sets of technological equipment and tooling, in turn included in autonomous ecosystems and infrastructure of special technological equipment as well as smart equipment.
Specifics of the current standards, conditions and requirements for the creation of design elements of the smart infrastructure complex and ecosystem of the smart object of confectionery production and special technological equipment, taking into account the solution of on-line in real time control and management tasks, as well as partial energy supply through the use of solar energy and, in addition, the creation of conditions and characteristics that allow the use of visual stabilisers of the psychological climate in production systems and premises.

Table of Contents

Accession :

Modern multidisciplinary digital technologies of confectionery production with elements of artificial intelligence and artificial neural networks, psychological nuances and aspects of marketing confectionery products created on the basis of and in the development of these technologies in their interaction with TRIZ.

Figure 1, - the figure shows a device for on-line mixing and simultaneous homogenisation of liquid fractions of confectionery production with formation of encapsulated structures before injection into heating chambers and thermodynamic objects of special technological equipment of confectionery production.

Changing the rules and criteria of industrial design in relation to advanced smart confectionery production

The introduction of new so-called smart technologies, the use of new so-called smart materials and composites, the replacement of traditionally accepted production methods with unusual ones, which help and are a prerequisite for a technological leap or breakthrough, increasing the efficiency of production, which begins to meet the criteria and characteristics of smart production, is today called a complex innovation process.

This process in conditions of different technical and technological cultures, in conditions of different level of starting positions for the beginning of the initiated process of innovations, may differ significantly, but the urgent need for the beginning of such a process exists and this fact does not cause any doubts.

In recent years, economies in virtually all industrialised countries have adopted and continue to adopt an increasingly pronounced innovative character.

And if at the beginning of this process the innovative breakthrough had a local significance and was observed in the field of high technologies, microelectronics and so-called ultra-precise digital technologies, then today the innovative process is becoming more and more focused on classical, basic technologies, energy, medicine, transport, including modular complexes of smart confectionery innovative technologies, i.e. covering all fundamental spheres of human life.

In order to increase the competitiveness of their confectionery products and technologies, entrepreneurs have to constantly search for new ways to increase efficiency, reduce energy consumption and direct energy costs, increase environmental safety and economic sustainability within each individual enterprise or company producing innovative and integrative confectionery products within the modern ecosystem of confectionery production and within the independent process of creation and preparation of production.

Figure 2 , - the figure also shows a device for on-line mixing and simultaneously linear homogenisation with the formation of encapsulated structures, if necessary, before feeding into the heating and thermal treatment chambers of thermodynamic objects as part of the technological equipment of confectionery production.

The device has 9 inlets and can simultaneously mix and homogenise 9 components of the mixture; this structure has not been used before, partly because there was no specific need for so many components in a confectionery mixture

New possibilities in designing and validating the performance of technical solutions also add elements of compositional design solutions and they become the main criteria for industrial design tools and methodology

A new look at the durability of a new product

Not so long ago, the durability of a product was one of the most important criteria for determining its commercial value; nowadays, with the time from the beginning of the marketing period of a new product to the beginning of the marketing period of an even newer product constantly decreasing, this time period is so short that it often makes no sense in the innovation process to concentrate attention and expend effort and money on an excessive increase in durability, which is longer than the period between the start of use of an existing confectionery product and the beginning of the marketing period of a new product.

Since this period may vary significantly for different product types, the concept of durability may be blurred over time and as a goal of the invention is not critical

There is one more subjective factor of durability that should be taken into account; based on the stereotypes of durability of different types of products, many commercial factors are determined, including the number of products demanded and therefore sold and their real price;

Imagine that a technical solution has been found to increase the durability of the product and then this factor reduces the quantity of the product required, while maintaining the current level of price that consumers are willing to pay for this product;

This leads to a decrease in sales of the companies producing the product and puts these companies before a choice - to agree with the innovation or to do everything to block the implementation and introduction of the innovation; as practice shows, these companies choose the second option and block the innovation and in this process the only loser is the inventor, who invented something rejected by the market or not fully understood by market players

A new look at the reliability of a new product

The issue of new product reliability and new criteria for assessing and calculating reliability are also undergoing fundamental changes; first of all, it is the relationship between reliability and the warranty obligations of the manufacturer of a new product to the consumer;

Very often the cost of warranty fulfilment is comparable to the cost of the product itself;

That is, reliability is a factor which, as one of the objectives of the invention, can determine (naturally in combination with other technical and operational factors achieved as a result of the invention) the level of commercial success.

In this case, the subjective factor of time also plays an important role, and more than necessary reliability can become a negative factor and play a cruel joke with the inventor, in a situation when the created ultra-reliable product turns out to be commercially unprofitable for the manufacturer;

New opportunities in the efficiency of system search and novelty analysis of previous technical solutions

It is clear that new information technologies open up new opportunities in the system search for similar solutions in the development of the technical solution under development

Let's imagine that during the preliminary design of the composition structure we came to the necessity to combine and integrate several classical solutions and new, let's say, digital technologies linking them into the composition, this is so to say at one level of horizontal integration and after that we also came to the necessity to reach the next level of integration with the inclusion of algorithms, software products and interfaces in the composition for communication with the previous level of integration.

At this stage of searching and analysing the search results, it is extremely important today to determine the predicted impact on the overall level of the product under study of the methods of using elements of artificial intelligence and artificial neural networks in its system structure

How to search, in what directions, and how to conduct this search most effectively and identify existing analogues of the composition to be created

In this case it seems to the author in this case the most probable for the beginning of search, after forming the composition of the created composition, to start system search, after decomposition and revealing of non-obvious, independent and independent technical solutions included in the composition.

After this stage of search it is better to choose one basic technical solution from these technical solutions and having made search on it, start to join other technical solutions included in the composition to the basic solution and conduct consecutive search on the basic technical solution with each joined technical solution and so on till achievement of full composition of the composition with the prospect of achievement by it of ideal final result

New possibilities for assessing the usefulness and feasibility (as well as the advisability) of optimising, modifying and upgrading known technical solutions in the confectionery industry.

Very often the new, is the well forgotten old.....

Therefore, when setting a problem and making a decision to start an innovative process of synthesis of a new confectionery product with non-obvious parameters and characteristics, it is desirable to check whether some functional elements of the composition to be invented have not been invented earlier.
If such or an equivalent solution is found, it is possible to replace materials, use new components and introduce a digital control and monitoring system into the future composition, which will allow to create a new technological composition that has the potential to be integrated into a composition of a higher level and retain all the properties and features of non-obviousness.

The impossibility of successful commercialisation of confectionery products without the formation of the principles of compositional construction of the schematic hierarchy and compositional structure of the ecosystem of a new technical solution.
As practice shows, the possibility of selling or licensing autonomous technical solutions, if they are not tied in advance to systems or solutions of a higher technological and qualitative level is reduced to zero.

Inventions, which are compositional in nature, in which there is at least a schematic solution for integration into technological - constructive - digital systems of a higher functional level, are implemented more confidently and in a shorter period of time, because the methodology and technique of this integration investors, buyers or consumers of the licence have in the description and in the formula of this integrative and compositional technical solution.

Suggested techniques and methods of composition style formation in the creation of new innovative solutions.

Thus, the compositional principles of technical solution , is a design and technological style of working out new technical solutions for their incorporation into existing technological schemes and configurations, including today and in smart machines and confectionery production;

Since the methodology of such incorporation can often be non-obvious, unique and possess substantial novelty, the description and claims having a compositional character, multi-level architecture of construction of causal links between the components of the composition and integrated into the design and technological links of the composition, distinctive features, determine to a large extent the commercial success of these innovations, especially in confectionery production;

Certainly today, the proven ability to effectively apply elements of artificial intelligence and artificial neural networks, creates additional interest in commercialisation, given all the additional benefits of such applications;

Techniques and methods of transition from the created composite foundation of a new technical solution to the basic basis of integrative invention of a new confectionery product

Design-technological composition in the field of formation and production of confectionery products in many cases requires additional, often fundamentally new links between components and elements of the composition, in other words, often having clearly expressed by properties and composition of design-technological composition, in order to turn it into a finished innovative product or product, it is necessary to find versions of integration of the composition in this final multiple and many levels and structural schemes integrated confectionery products.

There are many different versions of integration, the only important thing is that the end result of integration is a significant impact or quality leap that has no precedent and is not obvious from previous experience within the confectionery ecosystem.

Effect of claim limitations on the ability to reliably protect composite technical solutions in the confectionery industry

Limiting the number of claims in principle makes it difficult to reliably protect the invention, but a properly found principle of compositional structure at all levels of the hierarchy may conversely increase the degree and level of protection;

An ideal case is a system of cause-and-effect relations that allows to obtain the declared non-obvious effect only in the proposed system of compositional interrelations with clearly expressed conditions and features that determine the composition of the composition and independent functions of each of the elements of the obtained composition;

Due to the limited space and form, it makes sense to highlight in the compositional solution only those features and interrelationships that do not affect the known independent features and functions of each of the elements and components of the composition, but arose precisely as a result of the formation of the composition from the spheres of functional influence of the components of the composition on each other;

It can be said that in correctly selected components of the composition, when they are subordinated within the composition to the conditions and properties of the newly created technological system, a new non-obvious integrated system of features, interrelations, feedbacks and functions, possible only within the framework of this composition and, moreover, having a tendency to develop and improve intra-compositional relations, emerges;

Following these principles allows, within a limited number of claims, to focus attention only on the main distinctive features inherent in the composition, while ensuring the maximum level of protection of the composition and the maximum level of comprehensibility of the essence of the invention, despite its non-obviousness.

Proposed structure of an independent claim based on a composite technical solution

As the author has determined as a result of her first experiments and as recommended by those skilled in the art, an independent claim, where the invention is a complex confectionery composition, should have at least three main parts;

The first part carries the formulation of the commercial essence of the compositional invention of the confectionery product and its manufacture and should disclose the meaning and necessity of compositional integration for:

- a clear statement of the compositional problem
- limiting the degree of functional relationships in the composition and revealing the degree of necessity of each component of the composition for its formation and normal and effective functioning
- formulation of the name of the composition of a complex confectionery product

The second limiting part carries all the basic information about the invention as such and includes a characterisation of all the basic technical solutions inherent in the components of the composition and, at the claim drafting stage, does not qualify the presence of elements of substantial novelty in all commercial and design and technological aspects and relationships of the composition;

The third distinctive part carries information about components, their interrelations, materials, integrative elements and associated software products and their basic algorithms, each of which independently or in any combinations creates elements of essential novelty for a compositional, multiple and multi-level - integrated technical solution to create a fundamentally new confectionery product.

Proposed structure of a dependent claim based on a composite technical solution for the creation of a smart confectionery product and its production

Limiting the number of claims imposes a specific mission on each claim in the overall strategy of formulating and protecting the novelty of the invention and its non-obviousness;

On this basis, the second limiting part of such a claim should carry all the basic local and specific information about the invention as such and include a local and specific characterisation of all the basic technical solutions inherent in the components of the composition and, at the stage of drafting the claims, not qualified by the presence of elements of substantial novelty in all formulation and design and technological aspects and relationships of the composition, but having a clear definition of the necessary and compelling elements of the invention.

The third distinctive part carries local, specifically oriented on technical details and elements, information about components, their interrelations, materials, integrative elements and associated software products and their basic algorithms, each of which in local scale, independently or in any combinations creates local specific elements of essential novelty for non-obvious, compositional, repeatedly and on many levels - integrated technical solution of confectionery product.

Proposed methodology for designing prototypes of composite technical solutions of confectionery product and production, allowing to test and verify the correctness of each innovative element of confectionery product composition

Designing a prototype or a prototype sample of an innovative product - composition is most convenient when using a design programme and its engineering analytical applications; the author prefers to use the Solid Works software product, as this tool allows to build a working model of a composite technical solution and conduct a control simulation of its working cycle without resorting to expensive design and manufacturing of prototypes.

Analysis of found analogues and prototypes of the new composite technical solution

If as a result of the search will be found homogeneous, basic technical solutions in the first approximation are analogues or prototypes of the conceived composite technical solution of confectionery product, they should be tested in various variants and combinations of integration with additional elements and components of the composition under consideration;

If information about these prototypes or analogues is available in digital (digital) format, it is advisable to use Solid Works tools to build models of found technical solutions and conduct digital simulation of working cycles of these models to compare them with similar working cycles of the proposed invention - a composite technical solution of a smart confectionery product;

Examples of composite technical solutions that have been accepted by the market

An example of non-obvious composite technical solutions can be well known to everyone today products and products of computer technology, communication means, tablet computers and many other products of mass and non-mass demand;

Uncertainty in the wording of patent applications as the complex technical solutions underlying these products have led to and continue to lead to numerous patent disputes and wars;

A larger scale application of composite technical solutions and their integrated extensions and interpretations will reduce the heat of passion and possibly help to market innovative confectionery products;

There is an extremely important part of today's design process that corresponds to TRIZ technique number 40 used to achieve the perfect end result.

Reception 40. Application of composite materials

- move from homogeneous materials to composite materials.

It is well known that the substitution of one material for another is not recognised as an invention; the composite confectionery material itself is the subject or object of an original invention, but very often, when an ordinary construction material is changed to a composite, the properties and capabilities of the article or product are so changed that the article in which composites are used becomes completely new, unknown before, with completely new non-obvious functions and unusual technical and consumer characteristics;

Of course, in order to make such a change it is necessary to perform such a volume of work, which is comparable to the development of a fundamentally new technology or a fundamentally new smart confectionery product, and this is only possible for companies with powerful research departments;

To complete the basis for the analysis, we present information about the laws of development and formation of technical solutions known from TRIZ;

As of today, the so-called laws of development of technical and technological solutions are known, the correctness and accuracy of definitions and formulations of which are practically unproven;

Law of completeness of parts of the system

A prerequisite for the fundamental viability of a technical system is the presence and minimum operability of the main parts of the system.

Law of energy conductivity of the system

A prerequisite for the fundamental viability of a technical system is the through passage of energy through all parts of the system.

The law of harmonisation of the rhythm of the parts of the system

A necessary condition for the fundamental viability of a technical system is the coordination of rhythmics (frequency of oscillations, periodicity) of all parts of the system.

The law of increasing the degree of ideality of a system

The development of all systems is in the direction of increasing degrees of ideality.

Law of uneven development of parts of the system

The development of parts of the system is uneven. The more complex the system, the more uneven is the development of its parts.

Law of transition to the supersystem

Having exhausted the possibilities of development, the system is included in the supersystem as one of its parts. In this case, further development takes place at the level of the supersystem.

The law of transition from the macro to the micro level

The development of the working bodies of the system proceeds first at the macro level and then at the micro level.

The law of increasing the degree of real-field connections

The development of technical and commercial systems is moving in the direction of increasing the number of substance-field connections .

As practice has shown, unfortunately TRIZ and ARIZ are not adapted to the requirements and criteria of smart confectionery production and to the integrative use of elements of artificial intelligence and artificial neural networks, which does not quite correspond to the current situation in research and innovative design groups for a number of essential reasons;

Unfortunately, in the process of development of methods and equipment of system design, TRIZ and ARIZ have not been adapted to the real conditions and requirements of smart confectionery production not only from the technical and technological point of view, but also from the psychological point of view when approaching the issues of psychological preparation for the development of the marketing process and in general when building the commercialisation strategy and patent-licensing strategy of a smart confectionery product;

The transition to computer-aided design and engineering methods has drastically reduced the time for technical preparation of design and production of new innovative products, which determined a new psychological model of perception of this process, which did not provide a minimum psychological stimulus to any optimal use of TRIZ;

The use of computer modelling has dramatically reduced the number of errors and virtually eliminated the appearance on the market of unsuccessful solutions and confectionery products requiring correction and modernisation;

Therefore, in today's conditions it is psychologically more justified to completely replace outdated equipment and processes with new innovative ones, instead of gradual and limited modernisation in terms of efficiency and novelty.

Figure 3 , - the figure shows a view of a diesel generator, which is modernised by integrating into its fuel system a device for mixing and homogenising the fuel mixture - diesel fuel with methanol

Project development process for the integration of a device for dynamic mixing of fuel components in the fuel supply systems of thermal equipment (including boilers of all types).

Figure 4, - the figure shows a device for on-line mixing and homogenisation of liquid components of pastry mixtures and other culinary mixtures before injection or transfer to the thermal treatment chamber of a thermodynamic object of production equipment

Figure 5 , - the figure also shows a device for on-line mixing and homogenisation of fuel mixtures before injection into the combustion chamber of a thermodynamic object with internal parts

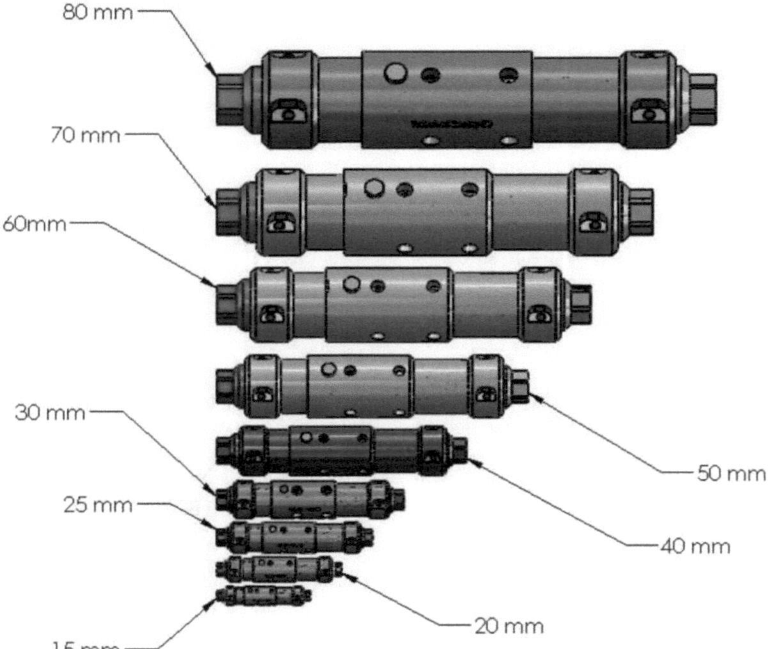

Figure 5 - 1, - the figure shows the size range of devices for on-line mixing and homogenisation of liquid confectionery components or fuel mixtures before injection into the combustion chamber of a thermodynamic object

All 9 sizes of devices for on-line mixing and homogenisation of pastry or fuel mixtures before injection into the combustion chamber of a thermodynamic object have an equivalent internal structure and the same design of internal parts.

Figure 6 , - the figure shows the instruments for monitoring the parameters of the obtained pastry or fuel mixture.

In the case of boilers, for example, the device should have a nozzle for a higher level of adaptation with fuel supply systems to the combustion chamber or burner.

Nozzle development is one of the integral parts of the project;

At the same time, the device itself can be presented as a stand-alone product or as a compact burner for thermal chambers in confectionery production;

This applies equally to boilers that use diesel as fuel as it does to boilers that use natural gas as fuel;

All variants and working versions of the device and its application technology can be developed at the same time, in parallel;

For devices designed for dynamic mixing of fuel components in which liquid fuel is dominant, there are a number of paradoxes that are inherent in all applications, including those for confectionery production;

The first paradox is characterised by the fact that in the pipeline, at the same cross-section, at the same initial pressure, at the same initial flow rate of the liquid non-compressible component, an annular vacuum zone is formed, without additional energy consumption

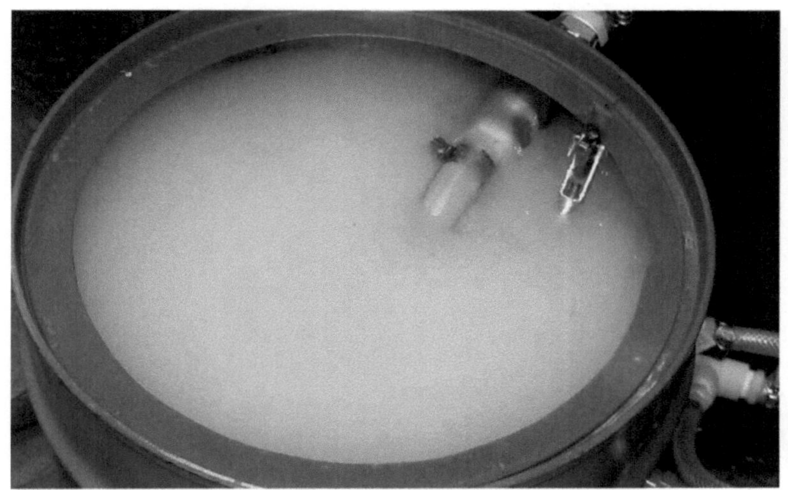

Figure 7 , - the figure shows, as an example, a view of the fuel mixture after structure recovery

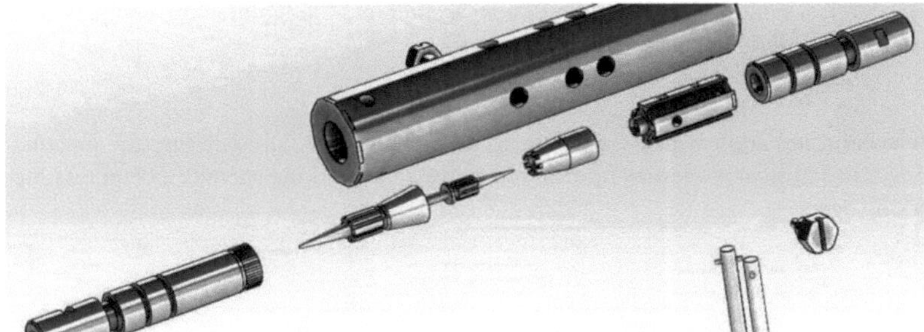

Figure 7 - 1 . - the figure shows the internal structure of the device

Figure 8. - the figure shows a view of the fuel mixture before structure recovery

This zone is a kind of boundary between an incompressible liquid and a compressible mixture of this liquid and a gas, in this case, air.

The second paradox is that within the same pipeline, the fluid entering the pipeline changes its physical properties from an incompressible working agent to a compressible working agent.

Figure 9 , - the figure shows systems using artificial intelligence and artificial neural networks used in qualification testing of technologies and devices.

The third paradox is that at the point where the flow changes its physical properties, the flows of the liquid and gaseous components are coaxial, with the flow of the liquid component covering the flow of the gaseous component.

Figure 10,- the figure also shows systems using artificial intelligence and artificial neural networks used in qualification testing of technologies and devices.

The directions of flow at the specified point are the same.

The fourth paradox is that there is a deep rarefaction or vacuum in the zone which is the boundary between the incompressible part of the flow and the compressible part of the flow, in conditions where two coaxial flows create, each, an annular rarefaction zone, of which one zone is created by the incompressible fluid flow and the second annular rarefaction zone is created by the compressible air flow.

Both of these zones are coaxial to each other and the flow thickness in them does not exceed 100 micrometres for liquid and 25 micrometres for gas.

The linear velocity in each of the streams in the rarefaction zone exceeds 100 metres per second, while no additional energy sources are used.

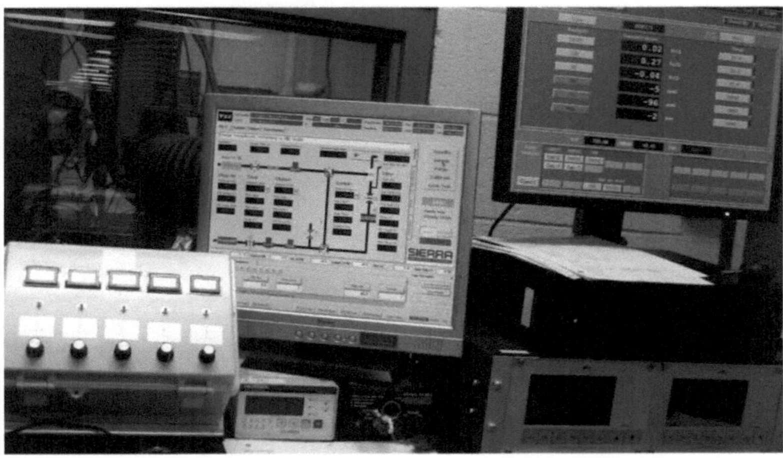

Figure 11, - the figure also shows fragments of the system using artificial intelligence and artificial neural networks used in confectionery production in qualification tests of technology and device.

The fifth paradox is that the mixture flow accumulates the kinetic energy of the flows of all components and the kinetic energy of the mixture leaving the device exceeds the kinetic energy of the liquid component of the mixture entering the device.

The sixth paradox is that under conditions of deep rarefaction and high linear velocity of flows, in the boundary zone separating the region of non-compressible liquid and compressible mixture, many capsules of confectionery liquid components or composite fuel are formed, more than 27 million spherical capsules with a diameter of no more than 50 micrometres per one litre of mixture or fuel composite.

In the specified capsules the core is a compressible element, - air, and the shell is a non-compressible element, - liquid or a homogeneous mixture of liquids.

A seventh paradox is that a liquid additional component of the fuel or confectionery composite, such as water, may be drawn into the boundary zone between the compressible and non-compressible portions of the flow, if necessary;

In this case, no additional energy is required to draw water into and mix with the liquid fuel component stream, but only the energy of the liquid fuel component stream.

Application of electromagnetic resonance method based on the principles of electromagnetic resonance spectroscopy for systemic and complex monitoring of biochemical processes related to the physiological activity of the human organism

Working principle of electromagnetic resonance sensor by electromagnetic resonance spectroscopy methods

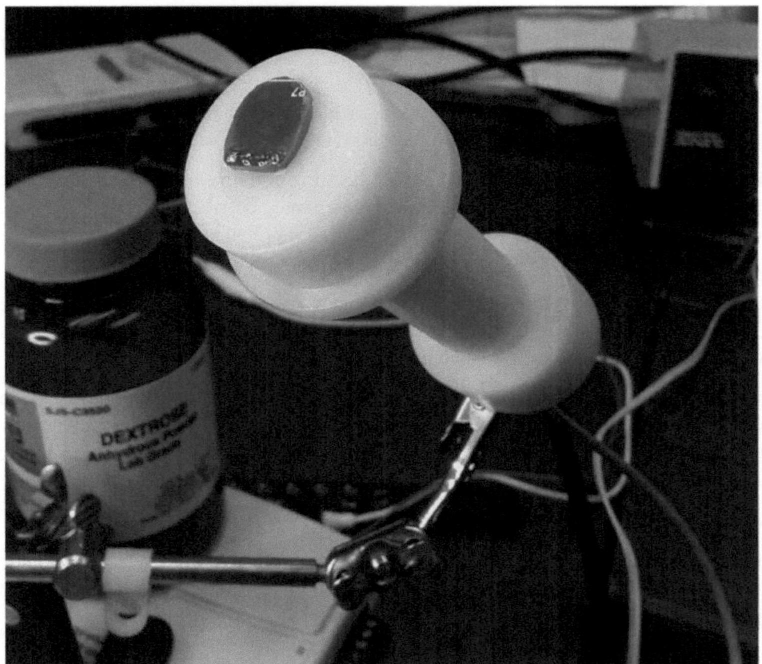

The method involves the creation of an alternating electro-magnetic field in the space in which the sample is placed.

This field mediates between the resonant circuit and the test specimen.

On the one hand, the resonant circuit is the emitter (radiator) of this field, and on the other hand, it is the acceptor (sensing element) of those changes in the electro-magnetic field that are introduced by the test sample.

Under the influence of an external alternating electro-magnetic field, such electrical phenomena as linear and eddy conduction currents, linear and eddy displacement currents, as well as linear and eddy ionic currents (ordered motion of ions) can be induced in the test sample, depending on its nature.

According to the principle of electromagnetic field formation, these electrical phenomena introduce distortions into the external alternating electro-magnetic field. These distortions are perceived by the solenoid of the electromagnetic resonance sensor (IR sensor).

A resonant circuit that includes this solenoid or its microelectronic equivalent changes its behaviour in the same way as if additional elements were added: a capacitor, an inductance and a resistor.
The sum of the additional capacitive, inductive and active resistances represents the additional impedance introduced into the system by the test sample, this attribute is what the IR sensor measures.

Changes in the parameters of the resonant circuit are reflected in changes in its complex amplitude and -frequency response , namely, the resonant frequency and amplitude of the circuit change.

By studying these changes, it is possible to judge about maximum active, reliably measurable and other indicative parameters of the investigated sample.

As it is clear from the above description of the operating principle of the electromagnetic resonance sensor according to the principles of electromagnetic resonance spectroscopy, the complex response from the measured object can be applied in all applications where eddy current test instruments (electromagnetic sensors) are used.

In addition, electromagnetic resonance sensors can be used in all applications where the measurement of characteristic parameters of an object placed between electrodes is used, e.g. in the case of a radio frequency impedance/material analyser.

The advantage that the IR sensor has over the above methods is the higher (at least ten times) sensitivity, which is provided by the use of the resonant circuit in its extremely reduced form, namely, consisting only of a coil of inductance or its microelectronic equivalent, which uses its intercurrent capacitance to create an oscillating circuit.

One- and two-part applications

This includes all applications in which it is sufficient to use a single electromagnetic resonant sensor that provides two parameters as an indication: resonant frequency and resonant amplitude.

These applications include all flaw detection applications and all applications that monitor changes in one or two parameters, such as chemical-mechanical and film and coating processes, object moisture determination, water salinity determination, etc.

Building an analytical system

For applications where changes in complex, multi-component and especially confectionery systems need to be measured and/or monitored, it becomes necessary to build an analytical system consisting of multiple sensors, each operating at its own unique operating frequency.

The initial data for construction of such system used for chemical and/or physical or complex integrative analysis of the investigated object are spectra obtained by methods of Electrochemical Electromagnetic Spectroscopy (EIS) and/or Dielectric Relaxation Spectroscopy (DRS) on measuring equipment used by these methods: radio frequency impedance/material analysers, pulse generators etc.

Reference samples of the investigated object with known variation of chemical components or with known variation of physical properties of the object are used to construct spectra.

The accuracy of measurements with the electromagnetic resonance analytical system will depend on the extent to which these reference samples fully cover all possible states of the object under investigation.

Then, using the chemo-metric* approach, the operating frequencies for electromagnetic resonant sensors included in the analytical system are determined on the basis of the obtained spectra.

The number of sensors in the system should be greater than or equal to the number of components being analysed.

The main criteria for selecting the operating frequency are the maximum change in impedance in accordance with the change in concentration of the component or physical property under investigation and the contrast of this response against the background of changes in other components or physical properties.

Proposed research design

Non-contact blood count monitoring (as an example)

This method involves installing a system of electromagnetic -resonance sensor modules along blood vessels in areas of the body where these vessels are closest to the body surface.

And the object of research can be different types of blood: arterial, venous and capillary.

To build such a system, a considerable amount of preparatory work must be done to find out how many sensors are needed and what operating frequencies they should have.

To do this there are at least two different approaches: synthetic and analytical.

Synthetic approach

This approach involves a gradual progression from simple measurements to more complex ones.

As measuring equipment it is possible to use the whole arsenal of devices intended for research in the field of Electrochemical Electromagnetic Spectroscopy (EIS), namely: pulse generators, impedance meters, etc.

The methodology of these studies consists in taking and analysing EIS spectra in the radio frequency range between 1-1000 MHz of various concentrations in distilled water of individual blood components and their combinations.

It is most appropriate to start with one-component solutions of dextrose and table salt.

The aim of the research is to identify those "promising" regions of the spectra where changes in the amplitude or phase of the EIS spectrum are more correlated with changes in the concentrations of the component to be measured (e.g.: dextrose).

Then solutions or suspensions of all other chemical and biological components of blood are investigated separately in order to select from the "promising" frequency ranges those areas where the influence of components not subject to investigation on the total electrochemical impedance is minimal.

Further for specification of working frequencies for electromagnetic resonance sensitive elements the construction and analysis of electromagnetic spectra are repeated, but not for one-component solutions, but for their mixture.

If the analysis of spectra of simple one-component solutions is possible "manually", then for the analysis of complex combinations of components one cannot do without the use of chemo-metric* methods.

Once the operating frequencies have been determined, a sensor system is constructed and calibrated using samples that cover all possible combinations of blood component concentrations.

The system is then tested on a human subject with parallel blood tests to determine the necessary correction factors to the calibration data to account for possible skin, blood vessel wall and muscle effects.

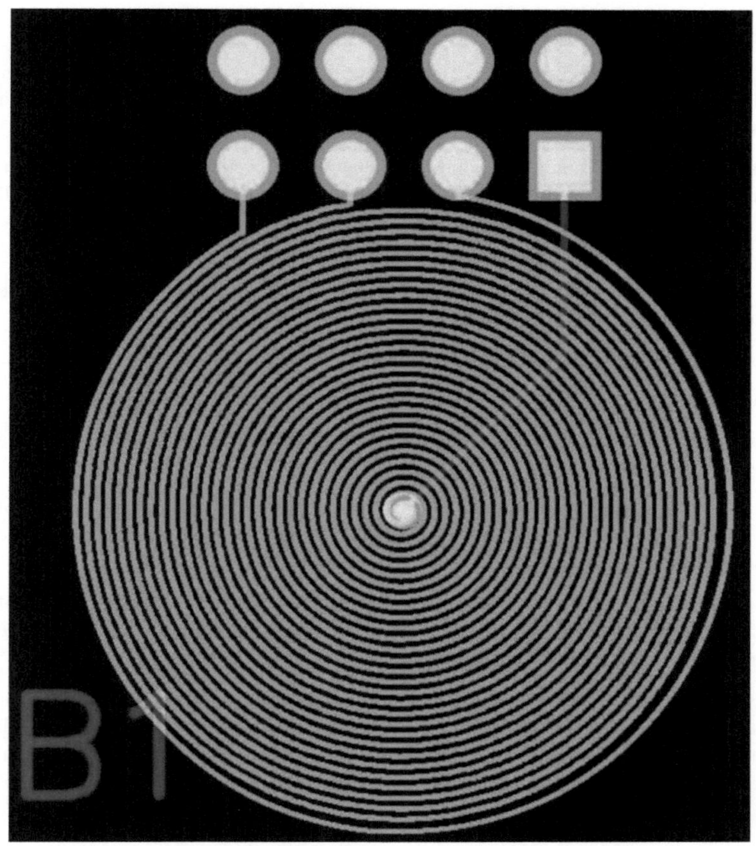

Analytical approach
This method involves periodic acquisition of EIS spectra by direct two-electrode impedance measurements of human body areas adjacent to blood vessels.

In parallel with taking spectra it is necessary to take a blood test and determine its component composition.

Next, by comparing changes in the EIS spectrum with changes in blood tests, using chemo-metrics* to determine the correlation coefficients between impedance and blood component concentrations at different frequencies.

By analysing the above mentioned correlation coefficients, the operating frequencies for the sensors are selected.

After construction of the system of electromagnetic resonance sensing elements, in order to calibrate it, it is necessary to make a series of measurements with parallel blood tests and determination of its component composition.

Non-contact monitoring of acupuncture points.

In all diseases, whether physical or mental, there are sensitive spots at certain points on the surface of the body, which disappear when the disease is cured.

These are so-called acupuncture points (the anatomical location of these acupuncture points can be seen in the "Atlas of Acupuncture"). These points are usually small nodules like fibrotic rheumatic nodules, often found on the back of the neck, on the shoulders or in the lumbar region.

But in many cases, instead of a nodule, it may be a band of muscle strain within the muscle group along with a particularly hard and compacted area.

These are often places that are slightly swollen or have discolouration.

Dr Voll created his method based on changes in conductivity at an acupuncture point.

His instrument is an ordinary ohmmeter operating on direct current, in other words at a frequency of zero hertz.

EIS technology can provide an order of magnitude more information about the state of the acupuncture point than the Voll method, and in the case of electromagnetic resonance sensitive elements, without direct electrical contact.

Nevertheless, to investigate changes in the electromagnetic characteristics of acupuncture points as a result of certain diseases it is possible to use a two-electrode scheme in combination with measuring devices;

After conducting comprehensive EIS studies and selecting working frequencies, it is possible to build a system of electromagnetic resonance sensitive elements that will monitor the state of human health and warn about all deviations from the normal functioning of his body.

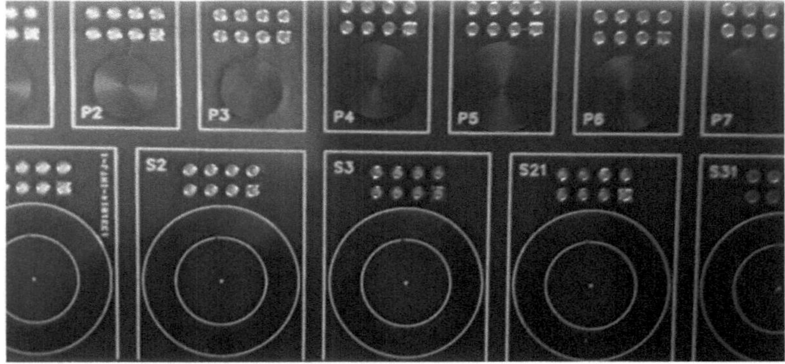

Schematic of planar Impedance Resonance Sensor connection

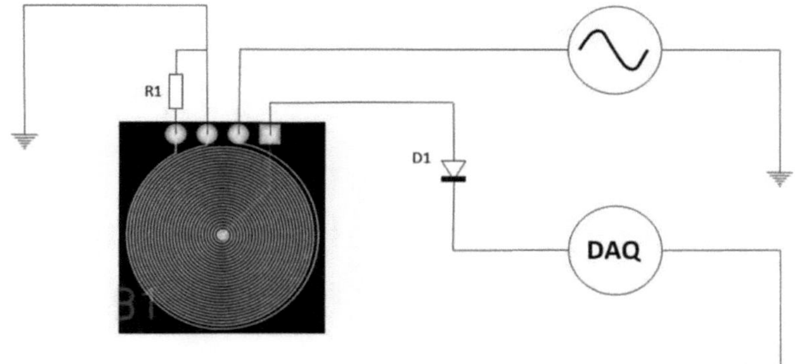

Excitation coil is connected to generator with frequency sweep. Resistance of R1 depends on output resistance of the generator and length of line between the sensor and the generator. If the length of line is less then ¼ of wavelength, the resistor can be omitted. Sensing coil is connected to Data Acquisition (DAQ). If operating frequency of the sensor is more than frequency capability of the DAQ, then the diode can solve that problem.

Both the generator and the DAQ are connected to personal computer and work under control of computer programme. The programme calculates resonant frequency and amplitude of the sensor and measure their changes caused by presence of analyte under test.

The feasibility of this type of monitoring has been little explored to date, but if successful, this project may prove to be the most feasible and effective.

Report on the implementation of the first and second stages of the project - Sensor module for non-contact control of glucose concentration level and its equivalents in blood using methods and devices characteristic for electromagnetic resonance spectroscopy.

Report on implementation of the first stage of the project - Sensor module for non-contact control of glucose and its equivalents concentration level in blood by means of methods and devices characteristic for electromagnetic resonance spectroscopy.

The first stage of work includes the following operations (10 items in total) :

- Defining the initial technical requirements for the sensor module
- Defining the initial specifications and characteristics of the sensor module
- Determination of the main distinguishing features and elements of fundamental novelty of the sensor module designs
- selection of the main basic invention underlying the design and usage parameters of the sensor module
- Determination of the main design and technological characteristics of the sensor module
- defining basic instructions and justifications for the order and methodology of application of the sensor module
- Determination of the main characteristics and profiles of construction materials used in assemblies and parts of the sensor module
- development of the programme and methodology of the stages of preliminary evaluation tests of the design and variants of its modifications depending on the current model of the sensor module
- development of initial technical requirements for special technological equipment for the implementation of the programme and methodology of the stages of preliminary evaluation tests of the design and variants of its modifications depending on the current model of the sensor module
- development of initial technical requirements for special laboratory control and analytical equipment for implementation of the programme and methodology of the stages of preliminary evaluation tests of the design and variants of its modifications depending on the current model of the sensor module

Labour intensity of the implementation of the first stage of work (the work was done using full time of two employees with total labour intensity and average cost of 1 working hour for each employee = 50 USD):

- definition and formulation of initial technical requirements for the sensor module; conducting a search for a depth of at least 20 years in order to qualify and confirm the degree of novelty of the main technical solutions required in the initial technical requirements; comparison and coordination of the parameters of the initial technical requirements with the requirements and limitations of current standards (6 working hours for each employee).

- definition of initial specifications and characteristics of the sensor module; search for at least 20 years in depth to qualify and confirm the degree of novelty of the main technical solutions required in the initial specifications; comparison and coordination of the parameters of the initial specifications with the requirements and limitations of the current standards (6 working hours for each employee)

- determination of the main distinctive features and elements of fundamental novelty of the sensor module designs; preliminary development of patent and licence strategy parameters for the basic sensor module design and for all associated assemblies and parts; conducting a search for a depth of at least 20 years in order to qualify and confirm the degree of novelty of the main technical solutions;

Qualification and formulation of the types of sensor modules and their designs, - as a cylindrical body with a flat coil on the working end; as a cylindrical body with a combined sensor in the form of a flat coil and as a combination of a flat coil on the end with a standard coil perpendicular to the flat coil on the end; as a standard coil transformed into a ring worn on the patient's finger; (9 working hours for each employee);

- Selection of the main basic invention, which is the basis for the design and parameters of the sensor module use; Conducting a search for a depth of at least 20 years in order to qualify and confirm the degree of novelty of the main technical solutions; Preparation of documents for the transfer of the invention and all international documents into the full ownership of the company - technology developer (3 working hours for each employee);
- Determination of the main design and technological characteristics of the sensor module ;

conducting a depth search of at least 20 years in order to qualify and confirm the degree of novelty of the main technical solutions; additional modelling of the possibility of installing a minimised sensor module on the basis of a flat coil
- micro-assemblies on the contact surface of smart watches and mobile phones (2 working hours for each employee);

- Definition of basic instructions and justifications for the order and methodology of the sensor module ;

search for at least 20 years in order to qualify and confirm the degree of novelty of the main technical solutions, including original solutions for the application of so-called smart technologies and devices; definition and qualification of new directions of implementation and use of various design options of sensor module for on-line control of various parameters in current liquids and aerosols with the use of elements of artificial intelligence and artificial neural networks; (4 working hours for each employee);

- determination of the main characteristics and profiles of structural materials and original coatings used in the assemblies and parts of the sensor module; development of the technique of coating the flat sensor coil with an artificial diamond film (10 microns thick); verification and analytical processing of the results of testing the coating of the surface of the flat sensor coil with an irradiated polyethylene film; conducting a search for a depth of at least 20 years in order to qualify and confirm the degree of novelty of the main technical solutions, including original solutions.

- development of the programme and methodology of the stages of preliminary evaluation tests of the design and variants of its changes depending on the current model of the sensor module; conducting a search for a depth of at least 20 years in order to qualify and confirm the degree of novelty of the main technical solutions, including original solutions for the application of so-called smart technologies and devices (3 working hours for each employee).

- development of initial technical requirements for special technological equipment for the implementation of the programme and methodology of the stages of preliminary evaluation tests of the design and variants of its changes depending on the current model of the sensor module; conducting a search for a depth of at least 20 years in order to qualify and confirm the degree of novelty of the main technical solutions required in the initial technical requirements for special technological and laboratory equipment; comparison and coordination of the parameters of the initial technical requirements for special technological and laboratory equipment; comparison and coordination of the parameters of the initial technical requirements for special technological and laboratory equipment.

- development of initial technical requirements for special laboratory control and analytical equipment for the implementation of the programme and methodology of the stages of preliminary evaluation tests of the design and variants of its changes depending on the current model of the sensor module; conducting a search for a depth of at least 20 years in order to qualify and confirm the degree of novelty of the main technical solutions required in the initial technical specifications; comparison and coordination of the parameters of the initial technical requirements and the initial technical specifications.

The total minimum labour intensity of all works on the first stage, without taking into account depreciation and other social charges and without taking into account related costs of all kinds, was 100 hours (one full working week for each employee, which in total corresponds to 5000 USD transferred as an advance payment for this stage).

Stage 1

Development of design and technological documentation for prototype production

Terms of reference for development

Harmonisation of the technical specification with the manufacturer

Initial technical requirements for the product

Product specifications

Harmonisation of technical specifications and technical requirements with the manufacturer

Development and coordination of technological instructions for the manufacturer

Report on implementation of the second stage of the project - Sensor module for non-contact control of glucose and its equivalents concentration level in blood by means of methods and devices characteristic for electromagnetic resonance spectroscopy.

Stage 2

Preliminary design

Technical project

Working Draft

Development of models for CNC machines

Transfer of documentation to the prototype manufacturer

Approval of manufacturing technology and acceptance inspection with the manufacturer

Development and agreement with the manufacturer of the programme and methodology of preliminary acceptance tests

ANNEX - 0 1

Methodology for selecting sensor operating frequencies

This methodology describes the first step required to build a system for monitoring the concentration of the components of the mixture under study.

By a mixture is meant any set of components, one of which is predominant in volume and is regarded as the conditional solvent and the others are regarded as conditional dissolved components.

The concept of a conditional solvent is not limited to a liquid one-component substance capable of forming solutions with other substances; it can also be a mixture of either gaseous, liquid or even solid substances (example - compound), which can serve as a base for the formation of not only solutions, but also suspensions, foam mists.

Accordingly, conditional dissolved components may not only be dissolved, but may also be present in solutions as suspensions, bacterial colonies, mist droplets, gas bubbles in foam, or filler particles in compounds.

Finding the optimum sensor operating frequency to build a single component mixture monitoring system.

Such a monitoring system can be used in processes where the concentration of one component may change while the concentrations of other components remain unchanged.

Preparation of samples for measurement
Two samples should be prepared with concentrations of the investigated component corresponding to the limits of the expected range of variation of this concentration.

Scanning
Using pulse generators scan the prepared samples by frequency, using the entire bandwidth of the pulse generator (in our case: from 0.100 MHz to 170 MHz).

During the scanning process, the pulse generator readings are taken and recorded in the form of changes in amplitude and phase shift of the current flowing through the sample relative to a harmonically varying probing voltage with stabilised constant amplitude.

Analysing the results

According to the scanning results it is necessary to select several frequencies at which the difference between the amplitudes of the tested samples reaches the highest values and several frequencies at which the difference in phase shifts reaches the highest values.

The selected frequencies will be the input data for designing the fabrication of trial resonant sensors.

Selecting the optimum sensor

The selection of a set of frequencies based on the results of scanning samples using a pulse generator is preliminary.

To decide on the optimal operating frequency of a single-component concentration monitoring system, it is necessary to test each trial resonance sensor, but no longer with two, but with at least 10 samples with different concentrations of the investigated component within the expected range of its variation.

After testing all prototypes of sensors, the best one can be selected, and preference should be given to sensors that, along with good sensitivity, have a monotonic change in readings in accordance with changes in the concentration of the controlled component (for ease of subsequent calibration).

However, lower frequencies are preferred for noise immunity. There are also design constraints to consider when selecting a sensor

Search for optimal sensor operating frequencies to build a two-component mixture monitoring system.

Such a monitoring system can be applied in technological processes in which the concentration of two components may change while the concentrations of other components remain unchanged.

Preparation of samples for measurement

It is necessary to prepare two samples for each investigated component with concentrations corresponding to the limits of the expected range of variation of these concentrations, as well as a sample in which these components are completely absent.

Scanning

Using a pulse generator scan the prepared samples by frequency, using the entire bandwidth of the pulse generator (in our case: from 0.100 MHz to 170 MHz).

During the scanning process, the pulse generator readings are taken and recorded in the form of changes in amplitude and phase shift of the current flowing through the sample relative to a harmonically varying probing voltage with stabilised constant amplitude.

Analysing the results

For each investigated component, according to the scanning results, it is necessary to select several frequencies at which the difference between the amplitudes of the investigated samples reaches the highest values, several frequencies at which the difference in phase shifts reaches the highest values and frequencies (or frequency ranges) at which there is no sensitivity to one component, and to the other there is.

Analyse the resulting frequency samples.

Depending on the results of the comparison, there are several possible algorithms for selecting the operating frequencies of the sensors.

A variant where there are frequencies where there is sensitivity to only one component.

This option is the most preferable for building a system for monitoring the concentrations of the investigated components.

If such frequencies exist for both one component and the other, the choice of operating frequencies is obvious:

For prototyping of resonant sensors, it is necessary to select such operating frequencies at which, in the absence of sensitivity to one component, the sensitivity to the other component is maximised.

At least one of these frequencies should be selected for each component.

If such frequencies exist only for one of the components, then for manufacturing of prototypes of resonant sensors for this component it is necessary to choose such operating frequencies, at which at absence of sensitivity to one component, sensitivity to the other is maximum, for the other component from its set of frequencies it is necessary to choose such operating frequencies, at which the difference in sensitivity to the investigated components is the greatest.

A variant where there are no frequencies at which sensitivity to only one component is present, but the resulting frequency samples do not match each other.

In this case for manufacturing of prototypes of resonant sensors it is necessary to choose from each set of frequencies such operating frequencies at which the difference in sensitivity to the investigated components is the greatest.

A variant where the obtained frequency samples are the same as each other.

This option is the most difficult to build a system for monitoring the concentrations of the investigated components.

If the set of frequencies for one component is exactly the same as the set for the other, it is necessary to check whether the proportion between the changes in amplitude or phase shift for one component and the changes in amplitude or phase shift for one component is maintained at all frequencies.

If everything matches, you should try to repeat the scan using other values of probing voltage.

If no differences can be achieved, it is likely that the components under investigation are indistinguishable in terms of electrochemical electromagnetic spectroscopy.

Nevertheless, even in this case it is possible to try to produce several prototypes of sensors with different operating frequencies corresponding to the highest sensitivity to changes in concentrations of the investigated components, since the resonant sensor has a more complex effect (magnetic field effect is added) on the investigated sample than the effect of the pulse generator

If testing of these sensors at least at one of the frequencies shows the presence of a change of proportion in sensitivity to changes in concentrations, then there is a fundamental possibility of building a system for monitoring the concentration of the investigated components, and the selectivity of this system will be the higher, the greater the difference in proportion.

Selection of optimum sensors

The selection of a set of frequencies based on the results of scanning samples using a pulse generator is preliminary.

To decide on the optimum operating frequencies of a two-component concentration monitoring system, it is necessary to test each trial resonant sensor, but no longer with two, but with at least 10 samples with different concentrations of the investigated components within the expected range of their variation.

After all prototype sensors have been tested, the best pair can be selected, with preference given to sensors that, in addition to good sensitivity, have a monotonic change in readings in accordance with changes in the concentration of the monitored components (for ease of subsequent calibration).

In this case, lower frequencies are the most preferable from the point of view of interference immunity.

Design constraints must also be considered when selecting sensors

Appendix - 0 2

Complex studies of the influence of the distance from the radiating element to the controlled liquid volume on the accuracy and sensitivity of the sensor module readings

As an active object of research are presented variants of flat coils in the form of printed circuit boards with a special topology that simulates the operation of a sensor module installed in a mobile cylinder Test and Measurement Systems.

In this variant the flat coil is installed in the bore of the housing of the module of non-contact control and measurement on the basis of electromagnetic resonance spectroscopy and for isolation is separated from the controlled liquid by a membrane made of irradiated polyethylene.

For the tests 6 dimensional types of membranes were prepared, and in each dimensional type there are 6 types of geometrical profile of the membrane cross-section.

After a full cycle of comparative tests, the most effective membrane with a rectangular cross-section, thickness - 0.5 millimetres, was determined to be the most effective in all parameters.

All further measurements and comparative conclusions were carried out on the module housing with this type of membrane

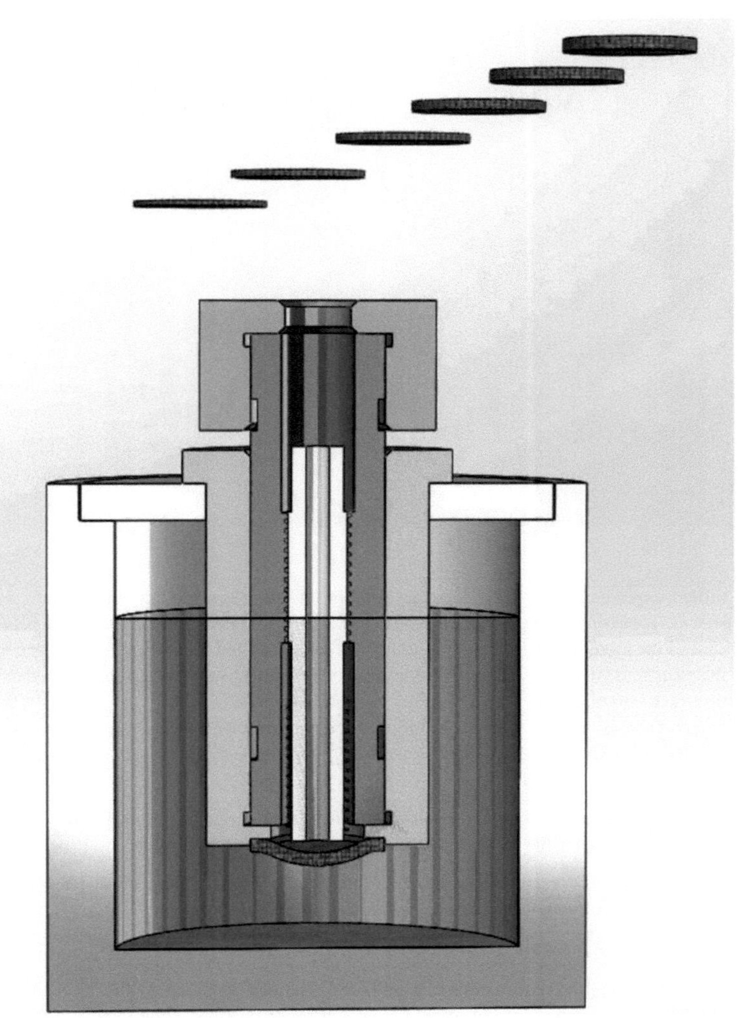

For further experiments we adopt a design solution of the sensor module with direct contact between the plane of the printed circuit board - flat coil, at which the effective distance between the plane of the topological pattern of the printed circuit board and the surface of the signal input into the measured volume is equal to the thickness of the protective coating of the printed circuit board surface - 50 microns.
The diagram shows the maximum sensitivity and accuracy of a sensor based on a flat coil, which is separated from the monitored object only by the thickness of the protective coating of 50 microns.

As can be seen from the diagram, the level of energy saturation of the signal and resonant reflection of the signal is the most intense with this method of arrangement.

In this case, the edge effect along the outer diameter of the printed circuit board - flat coil does not affect the power characteristics of pulses and their return derivatives.

In case of manufacturing of the printed circuit board - flat coil by methods of RITM technology and obtaining the thickness of the board only equal to 50 microns, the influence of the edge effect and its destructive phenomena are reduced to zero.

It should be noted that the distance between the radiating surface of the flat coil and the plane limiting the upper level of the volume to be measured (in the case of the present test, a controlled liquid with different concentrations of sugar content or its equivalents) does not depend on the fact that this distance is due to a membrane or that this distance is filled with air.

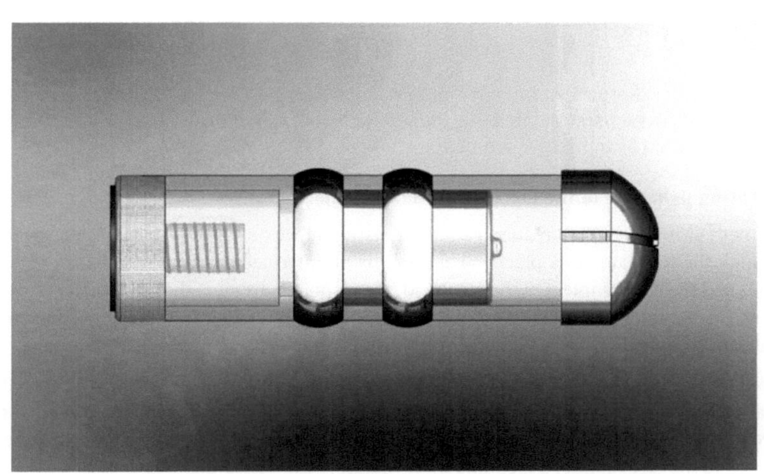

Appendix - 03

Application (examples) of electromagnetic resonance method for monitoring of biochemical processes related to physiological activity of the human body consuming smart innovative confectionery products

Operating principle of the electromagnetic resonance sensor

The method involves the creation of an alternating electro-magnetic field in the space in which the test sample is placed. This field is an intermediary between the resonant circuit and the test sample. On the one hand, the resonant circuit is the emitter (radiator) of this field and, on the other hand, the acceptor (sensing element) of those changes in the electro-magnetic field, which are introduced by the test sample.

Under the influence of an external alternating electro-magnetic field, such electrical phenomena as linear and eddy conduction currents, linear and eddy displacement currents, as well as linear and eddy ionic currents (ordered motion of ions) can be induced in the test sample, depending on its nature.

According to the principle of electromagnetic fields, these electrical phenomena introduce distortions into the external alternating electro-magnetic field.

These distortions are sensed by the solenoid of the electromagnetic -resonance sensor (IR sensor).

The resonant circuit, which includes this solenoid, changes its behaviour in the same way as if additional elements were added: a capacitor, an inductance and a resistor.

The sum of the additional capacitive, inductive and active resistances represents the additional impedance introduced into the system by the test sample, this attribute is what the IR sensor will measure.

Changes in the parameters of a resonant circuit are reflected in changes in its -frequency response , namely, the resonant frequency and amplitude of the circuit change.

By examining these changes, it is possible to judge the impedance of the sample under study.

As it is clear from the above description of the principle of operation of the electromagnetic resonance sensor, the complex response from the measured object can be applied in all applications where eddy current test apparatus is used .

In addition, electromagnetic resonance sensors can be used in all applications where impedance measurement of an object placed between electrodes is used, such as in the case of a radio frequency impedance/material analyser. The advantage that the IR sensor has over the above methods is the higher (at least ten times) sensitivity provided by the use of the resonant circuit in its extremely reduced form, namely consisting only of an inductance coil that uses its intercurrent capacitance to create an oscillating circuit.

One- and two-part applications

This includes all applications in which it is sufficient to use a single electromagnetic resonance sensor that provides two parameters as an indication: resonance frequency and resonance amplitude.

These applications include all flaw detection applications and all applications that monitor changes in one or two parameters, such as chemical and mechanical machining processes and film and coating processes, determining the moisture content of an object, determining the salinity of water, etc.

Building an analytical system

For applications where changes in complex, multi-component systems need to be measured and/or monitored, it becomes necessary to build an analytical system consisting of multiple sensors, each operating at its own unique operating frequency.

The initial data for construction of such system used for chemical and/or physical analysis of the investigated object are spectra obtained by methods of Electrochemical Electromagnetic Spectroscopy (EIS) and/or Dielectric Relaxation Spectroscopy (DRS) on the measuring equipment used by these methods: radio frequency impedance/material analysers, pulse generators, etc.

Reference samples of the investigated object with known variation of chemical components or with known variation of physical properties of the object are used to construct spectra.

The accuracy of measurements with the electromagnetic resonance analytical system will depend on the extent to which these reference samples fully cover all possible states of the object under investigation.

Then, using the chemo-metric* approach, the operating frequencies for electromagnetic resonance sensors included in the analytical system are determined on the basis of the obtained spectra.

The number of sensors in the system should be greater than or equal to the number of components being analysed.

The main criteria for selecting the operating frequency are the maximum change in impedance in accordance with the change in concentration of the component or physical property under investigation and the contrast of this response against the background of changes in other components or physical properties.

LIST OF REFERENCES, PATENT AND LICENCE INFORMATION

ANNEX 3

nited States Patent Application	**20180115003**
Kind Code	**A1**
REYTIER; Magali ; et al.	**April 26, 2018**

SOFC-BASED SYSTEM FOR GENERATING ELECTRICITY WITH CLOSED - LOOP CIRCULATION OF CARBONATED SPECIES

Abstract

A reversible SOFC-based system for generating electricity, including: a *solid-oxidefuel-cell* (SOFC) stack containing at least one elementary solid-oxide electrochemical cell, each of which is formed from a cathode, an anode and an electrolyte intermediate between the cathode and the anode; a separator of liquid and gas phases, which separator is connected to the outlet of the *fuel-cell* stack; a methanation reactor suitable for implementing a methanation reaction, the inlet of which is connected to the outlet of the phase separator and the outlet of which is connected to the inlet of the ***fuel-cell*** stack so that the mixture issued from the methanation reactor is introduced into the ***fuel-cell*** stack; and a tank for reversibly storing hydrogen, suitable for storing hydrogen, the outlet of which is connected to the inlet of the methanation reactor.

APPENDIX 4

nited States Patent Application	**20180202383**
Kind Code	**A1**
Alrefaai; Kutaiba ; et al.	**July 19, 2018**

METHODS AND SYSTEM FOR CENTRAL FUEL INJECTION

Abstract

Methods and systems are provided for leveraging the charge cooling effect of a manifold ***fuel*** injection. A charge cooling effect of a scheduled manifold ***fuel*** injection may be predicted based on feedback received from a manifold charge temperature sensor during a preceding manifold injection event. If sufficient charge cooling is not predicted, the manifold ***fuel*** injection is temporarily disabled.

APPENDIX 5

United States Patent Application	**20180266377**
Kind Code	**A1**
NOGUCHI; Koji	**September 20, 2018**

NOZZLE PLATE FOR FUEL INJECTION DEVICE

Abstract

A nozzle hole of a nozzle plate is coupled to a ***fuel*** injection port of a ***fuel*** injection device via a swirl chamber and first and second ***fuel*** guide channels opened into the swirl chamber. The swirl chamber is formed by combining first and second ellipticalshaped recessed portions. The first ***fuel*** guide channel opens at a side of a short axis of the first elliptical-shaped recessed portion and a side of the short axis that does not overlap with the second elliptical-shaped recessed portion, and the second ***fuel*** guide channel opens at a side of a short axis of the second elliptical-shaped recessed portion and a side of the short axis that does not overlap with the first elliptical-shaped recessed portion. The first and second ***fuel*** guide channels have depths deeper than those of the swirl chamber and extend inside of the swirl chamber while gradually reducing cross-sectional areas.

ANNEX 6

United States Patent Application	**20180274786**
Kind Code	**A1**
BAYA TODA; Hubert ; et al.	**September 27, 2018**

COMBUSTION CHAMBER OF A TURBINE, IN PARTICULAR A THERMODYNAMIC CYCLE TURBINE WITH RECUPERATOR, FOR PRODUCING ENERGY, IN PARTICULAR ELECTRICAL ENERGY

Abstract

A combustion chamber (18) of a thermodynamic cycle turbine with a recuperator, for electrical energy production, comprising a casing (56) housing a flame tube (64) with a perforated diffuser for passage of the hot compressed air, a primary zone (ZP) that receives part of the hot compressed air flow and where combustion takes place, and a dilution zone (ZD) where the burnt gases from the primary zone mix with the remaining part of the hot compressed air flow, said chamber further comprising an injection means (76) for injecting at least one ***fuel***. The flame tube carries a flame stabiliser (82) comprising a perforated diffuser (88), at least one combustion gas recirculation passage (98) and a ***mixing*** tube (94).

ANNEX 7

United States Patent Application	**20180320625**
Kind Codeм	**A1**
Surnilla ; Gopichandra ; et al.	**November 8, 2018**

METHODS AND SYSTEM FOR CENTRAL FUEL INJECTION

Abstract

Methods and systems are provided for adjusting engine operating conditions for mitigation of pre-ignition in one or more engine cylinders. In one example, a method may include, in response to indication of pre-ignition, manifold charge cooling may be increased by increasing the portion of ***fuel*** delivered to the engine via manifold injection relative to the portion to ***fuel*** delivered via one or more of port and direct injection, while maintaining engine operation at or around a stoichiometric ***airfuel*** ratio.

ANNEX 8

United States Patent Application	**20180328315**
Kind Code	**A1**
TANIEL; Roman	**November 15, 2018**

EMULSIFYING SYSTEM AND EMULSIFYING PROCESS

Abstract

What is proposed is an emulsifying system with an emulsifying device and an injection nozzle as well as as an emulsifying device for producing a water ***-fuel*** emulsion for an internal combustion engine, wherein the emulsifying device is embodied as a rotor-stator emulsifying device and/or fluid flow machine and/or is connected or connectable directly to an injection nozzle. The emulsifying device has a housing and a shaft, the shaft being drivable in a contactless manner, the housing having a guide apparatus having a plurality of guide channels for guiding the flow, and/or the housing being made at least partially from fibre composite material. Moreover, an emulsifying method for producing a ***water-fuel*** emulsion is proposed in which water and ***fuel*** are fed to a rotor-stator emulsifying device and/or fluid flow machine for producing the ***water-fuel*** emulsion, and/or in which water and ***fuel*** are premixed in a first emulsifying stage and fed via a guide apparatus having a plurality of guide channels to a second emulsifying stage.

ANNEX 9

United States Patent Application	**20180334968**
Kind Code	**A1**
Shirahashi; Naotoshi ; et al.	**November 22, 2018**

METHOD AND DEVICE FOR CONTROLLING FUEL INJECTION OF DIESEL ENGINE

Abstract

A method of controlling a ***fuel*** injection of a diesel engine for performing a plurality of ***fuel*** injections to cause a plurality of combustions inside a cylinder in one combustion cycle, is provided, which includes acquiring an oxygen concentration inside the cylinder, performing, on compression stroke, the plurality of ***offuel*** injections at substantially even injection intervals while increasing the injection intervals as the oxygen concentration decreases, and performing, after the plurality of ***fuel*** injections, another ***fuel*** injection including a larger injection amount than in the plurality of ***fuel*** injections, near a top dead centre of the compression stroke.

ANNEX 10

United States Patent Application	**20180340486**
Kind Code	**A1**
Marmorini; Luca	**November 29, 2018**

METHOD TO CONTROL THE COMBUSTION OF A COMPRESSION IGNITION INTERNAL COMBUSTION ENGINE WITH REACTIVITY CONTROL THROUGH THE INJECTION TEMPERATURE

Abstract

Method to control the combustion of a compression ignition engine with reactivity control through the injection temperature; the control method provides for the steps of: establishing a quantity ***offuel*** to be injected into a cylinder; injecting a first fraction of the quantity ***offuel*** fed by a first feed system without active heating devices, preferably equal to at least 70% of the quantity of ***fuel,*** at least partially during the intake and/or compression stroke; injecting a second fraction of the quantity ***offuel*** fed by a second feed system provided with at least one active heating device, and equal to the remaining fraction of the quantity of ***fuel,*** into the cylinder at the end of the compression stroke and preferably at no more than 60.degree. from the top dead centre; and heating the second fraction of the quantity of ***fuel*** to an injection temperature of over 100.degree. C., before injecting the second fraction of the quantity ***offuel.***

ANNEX 11

United States Patent Application **20190195183**
Kind Code **A1**
HASHIZUME; Takeshi **June 27, 2019**

INTERNAL COMBUSTION ENGINE

Abstract

An internal combustion engine includes a ***fuel*** injection nozzle in which a nozzle hole that injects ***fuel*** is provided to be exposed to a combustion chamber from a cylinder head of the internal combustion engine, and a hollow duct in which an inlet and an outlet are exposed to the combustion chamber. The duct is provided to penetrate through an inside of the cylinder head so that ***fuel*** spray injected from the nozzle hole of the ***fuel*** injection nozzle passes from the inlet to the outlet. The duct is preferably configured so that a direction from the inlet to the outlet corresponds to a direction of the ***fuel*** spray injected from the nozzle hole.

ANNEX 12

United States Patent Application **20200011236**
Kind Code **A1**
TANNO; Shiro ; et al. **January 9, 2020**

COMPRESSION-IGNITION INTERNAL COMBUSTION ENGINE

Abstract

A compression-ignition internal combustion engine including a ***fuel*** injection nozzle including a tip end portion exposed in a combustion chamber and a nozzle hole formed at the tip end portion; and a passage forming member forming a flow guide passage through which ***fuel*** injected from the nozzle hole passes. The passage forming member includes a passage wall portion located radially outward of the flow guide passage. The passage wall portion includes a first layer that is a base portion connected to a cylinder head, and a second layer located radially outward or radially inward of the first layer. The toughness of the first layer is higher than the toughness of the second layer. The thermal conductivity of the second layer is lower than the thermal conductivity of the first layer.

ANNEX 13

United States Patent Application	**20190323415**
Kind Code	**A1**
CORRIGAN; Daire James ; et al.	**October 24, 2019**

HIGH-PERFORMANCE INTERNAL COMBUSTION ENGINE WITH IMPROVED HANDLING OF EMISSION AND METHOD OF CONTROLLING SUCH ENGINE

Abstract

An internal combustion engine comprising: at least a cylinder; at least an intake valve acting on an intake port for controlling the airflow entering the cylinder; at least an injector for supplying uncombusted ***fuel*** to the cylinder; at least an outlet valve acting on a respective outlet port 18 for controlling the flow of the exhaust gases at the outlet of the cylinder; a piston sliding in a linear manner within the cylinder; at least a first spark plug arranged in a position adjacent to the injector and acting within the combustion chamber; a pre-chamber communicating with the combustion chamber; and a second spark plug acting within the pre-chamber; the first spark plug is arranged in an intermediate position between the pre-chamber and the injector.

ANNEX 14

nited States Patent Application	**20190277185**
Kind Code	**A1**
HASHIZUME; Takeshi	**September 12, 2019**

INTERNAL COMBUSTION ENGINE

Abstract

An internal combustion engine includes a ***fuel*** injection nozzle provided with a nozzle hole for injecting ***fuel,*** the nozzle hole exposed from a cylinder head of the internal combustion engine to a combustion chamber, and a hollow duct, an inlet and an outlet of which are exposed to the combustion chamber. The duct is provided in a manner allowing ***fuel*** spray injected from the nozzle hole of the ***fuel*** injection nozzle to pass through from the inlet to the outlet. The ***fuel*** injection nozzle and the duct are configured such that a part ***offuel*** spray that is injected in pilot injection that is performed before main injection directly adheres to an inner wall surface of the duct.

ANNEX 15

United States Patent Application **20200172822**
Kind Code **A1**
Asmatulu; Ramazan ; et al. **June 4, 2020**

WATER IN FUEL NANOEMULSION AND METHOD OF MAKING THE SAME

Abstract

A method of producing a nanoemulsion is disclosed that provides an oleaginous base ***fuel,*** and water in an amount of at least 10 wt %. A first nonionic surfactant, a second nonionic surfactant and a third nonionic surfactant are mixed in substantially equal weight ratios into a surfactant mixture. The surfactant mixture is mixed with the water and the base ***fuel*** to form the nanoemulsion ***fuel.*** A nanoemulsion ***fuel*** composition can comprise an external oleaginous phase comprised of base ***fuel,*** an internal aqueous phase comprised of water, and a surfactant mixture comprised of a plurality of surfactants. The first surfactant can be derived from ethylene oxide, the second surfactant and the third surfactant are detergents having a fatty acid.

ANNEX 16

United States Patent Application **20200182466**
Kind Code **A1**
Sadasivuni; Suresh **June 11, 2020**

PILOT BURNER ASSEMBLY WITH PILOT-AIR SUPPLY

Abstract

A pilot burner assembly for a combustion volume in a gas turbine engine includes a pilot burner, a ***pilot-fuel*** supply line, and a pilot-air supply line. The pilot burner has a burner face that includes a plurality of ***pilot-fuel*** injection holes. The ***pilotfuel*** injection holes provide a ***pilot-fuel*** to the combustion volume. The ***pilotfuel*** supply line is fluidly connected to the ***pilot-fuel*** injection holes for supplying the pilot-fuel to the ***pilot-fuel*** injection holes. The pilot-air supply line provides a pilot-air to the pilot burner. The pilot-air is supplied to the combustion volume through the burner face. Pilot-air injection holes are located on the burner face and fluidly connected to the pilot-air supply line. The pilot-air injection holes inject the pilot-air into the combustion volume. A gas turbine has the pilot burner assembly.

ANNEX 17

United States Patent Application	**20200200118**
Kind Code	**A1**
Matsuo; Takeru ; et al.	**June 25, 2020**

CONTROL DEVICE FOR COMPRESSION IGNITION ENGINE

Abstract

A control device for a compression ignition engine is provided, which causes an injector to perform a pre-injection and a main injection, sets ***fuel*** injection timings of these injections so that an interval between a first peak of a heat release rate resulting from the combustion ***offuel*** injected by the pre-injection and a second peak of the heat release rate resulting from the combustion of ***fuel*** injected by the main injection becomes an interval to make pressure waves caused by these combustions cancel each other out, and when an increase of an intake air temperature is detected, controls the injector to reduce the injection amount of the pre-injection and retard the injection timing of the pre-injection compared with a case where the increase of the intake air temperature is not detected under a condition that engine load and speed are the same.

ANNEX 18

Process of waste-free processing of chicken eggs.

All equipment is stainless steel, glass, enamel.

The prerequisite for a quality process is fresh eggs, cleanliness and careful temperature control at all stages of the process.

1. Egg powder production.

a)Preparing the eggs for the process.

Eggs for production come cooled (to 5-10 C) from the warehouse. They are weighed. Contaminated eggs are washed at ~32 C (washing&sanitising) and only then they go to the breaking machine. Spoilt eggs are discarded (check by inspection).

***It is extremely important to control the microbiological purity of eggs!

b)Separation of egg and shell contents.

Eggs are broken using breaking&separation machines.

in)The shell is transferred to a centrifuge where the liquid egg residue is separated from the shell. The shells in the centrifuge are thoroughly washed, transferred to a dryer and then ground in a mill. Eggshell powder (~94% $CaCO_3$) is used as a mineral supplement in poultry and livestock feed. Eggshell powder has a moisture content of ~1%.

d)The liquid eggs obtained in stages b) and c) are combined, filtered and transferred to a special container where they are cooled to 4 C and slowly mixed.

e)In stage b), both egg melange (yolk + albumen) and separated albumen and yolk can be obtained.

***** For egg white production, effective separation of egg white from yolk is extremely important. The yolk must not "contaminate" the albumen.**

Modern apparatus allows the production of protein containing no more than 0.02% yolk.

e)Pasteurisation.

From the storage tank (stage d), the egg melange/egg white/egg yolk is transferred by pump to the turbulent heater for pasteurisation.

Pasteurisation is carried out at 64-66 C for 2-4 min. This ensures deactivation of most microbes such as E. coli and Salmonella.

It is possible to pasteurise the liquid egg mixture/egg white with hydrogen peroxide at a reduced temperature of 7 -13 C to avoid possible changes in the composition of the egg mixture/egg white due to the temperature. Approximately 1.31 35% hydrogen peroxide per 1 tonne of liquid egg mixture is added slowly with continuous stirring. After a 20 min residence time, 100-150ml of Catalase enzyme (C641L) is added to the reactor to remove any residual hydrogen peroxide.

For pasteurisation of egg white, hydrogen peroxide is injected at a rate of 3ml/min for 10-16 hours.

w)Spray drying.

The warm liquid product from the pasteurisation stage is pumped from the pasteurisation stage into a tank, from which it is pumped by a high-pressure pump into the spray dryer in the form of a mist. The inlet air temperature varies between +150-+200 C and the outlet air temperature is +55-+65 C. The resulting powder settles in the dryer. The resulting powder settles on the conical walls of the dryer and is collected in the lower part of the dryer. After cooling down to room temperature, the powder is packed. The dried product has a moisture content of 2-4% and a density of 0.3-0.35 g/cm3.

Reducing the sugar content of egg powder/white.

Egg powder and egg whites change their colour to a brownish colour when pasteurised and dried. This is due to a reaction between glucose and protein at elevated temperatures - the Maillard reaction.

Glucose oxidase enzyme (G168L) 100-150ml per 1 tonne of liquid product is recommended to reduce the sugar content of egg powder/white. The enzyme/coenzyme system - Glucose oxidase/Catalase efficiently converts egg glucose (egg melange or egg white) into acid (gluconic acid).

Sugar purification of egg white by Propionibacterium shermanii (+37 C, 24 hours) results in complete removal of glucose, which is the major part of protein carbohydrates, enrichment of egg white with vitamin B12 and the preserving agent propionate.

Pre-fermentation of liquid egg raw material with baker's yeast can be used to significantly reduce the sugar content of egg powder/egg white. The glucose content of egg white is drastically reduced in the presence of 0.1% Saccharomyces cerevisiae, a baker's yeast extract.

Fermentation of whole eggs in the presence of 0.2-0.4% wet baker's yeast at 22-23 C takes 2-4 hours. Centrifugation of the sugar-free liquid removes yeast cells and improves the odour of the product. Acidifying the egg melange to a pH below 6.0 increases the fermentation rate. Optimum fermentation temperature +30 C, 0.07-0.15% yeast and fermentation time 2-3 hours.

What is left over after making egg powder? - Water and eggshells.

Egg powder is an alternative to fresh eggs, both in terms of ease of use and storage. The powder is mixed with water to produce liquid eggs, which can then be used like fresh eggs. From 1kg of egg powder, 4kg of liquid eggs can be prepared.

Egg powder can be used in most egg dishes or in recipes that call for the use of eggs.

Egg powder can be used in mixture with other powders of organic and inorganic origin.

The refining of egg powder involves, firstly, converting it into a solution and, after the refining process, drying it again. This is not advisable, both from an economic point of view and from the point of view of the additional possibility of losing part of the quality and quantity of the product. It is much more efficient to carry out purification (from sugar - glucose) even before spray drying.

2. Egg butter production.

The process involves the use of volatile organic solvents and therefore all equipment and communications must be made of chemically resistant materials (stainless steel, enamel, glass) and the equipment must be explosion-proof and fireproof.

Equipment

This process uses:

1. Two reactors with stirrers R1 and R2
2. Three containers for three solvents E1, E2 and E3 and four containers for collecting and storing finished products
3. Three capacitors
4. Flow-through centrifuge
5. Four evaporators I1, I2, I3 and I4
6. Vacuum rotary evaporator
7. Three intermediate tanks for three secondary solvents S1, S2 and S3
8. Solvent drying system
9. Pumps for pumping solvents, extracts and products
10. System for filling and packaging of products
11. Refrigerators for storing egg oil and lecithin

Production and storage rooms must be ventilated and meet the standards for food production.

PROCESS DESCRIPTION

The process of separating egg contents into their individual components is a series of periodic, sequential processes of extraction of substances, filtration/centrifugation of solutions and evaporation of solvents, carried out in closed vessels. The extraction is carried out under constant stirring. The evaporated solvents are condensed into separate containers and reused in the process. Over time, water accumulates in the solvents and they are then dried with suitable adsorbents, e.g. granulated aluminium oxide or CaC1

.2

Process 1. Separation of egg oil from egg powder.

In the reactor R1 (V=1m3) containing solvent S1 - CH2CI2 - di chlorine methane (500 litres) at a temperature of 20°C, at constant stirring through the top hatch is loaded egg powder 250kg. The content of the reactor is stirred for 1 hour, after which by means of a submersible pump with a filter-tip, the solution of egg oil in the solvent S1 from R1 is first fed to a flow centrifuge (to remove the remains of egg powder, caught in the solution), and then - to the vacuum evaporator I1. Here the egg oil is separated. The egg oil is collected at the bottom of the evaporator and drains into a separate container E4.

The solvent from the vacuum evaporator I1 is first condensed into the intermediate vessel ES1 and then pumped into its original vessel E1 for subsequent reuse in the process.

A fresh portion of solvent S1 is added to the remaining wet powder mass in reactor R1 and the process is repeated twice more. The powder residue in reactor R1 is combined with the precipitate collected on the centrifuge, washed again with solvent S1 and this suspension is fed to the centrifuge under stirring. The output solution is fed to the vacuum evaporator I1 to extract an additional portion of egg oil. The egg oil flows into the container E4 and the solvent into the intermediate container ES1 and is then pumped to its original container E1.

Depending on the fat content of the original egg powder (39-40%), up to 100kg of egg oil can be practically quantified (more than 95%) from 250kg.

Egg oil is a clear, yellow/yellow-red, oily liquid with a characteristic egg odour.

The main feature of this technology is the use of a solvent that does not mix with water, in which pathogens cannot develop and in which albumin is absolutely insoluble.

What is left over after egg butter is produced? Albumin. And the solvent is returned to the process after drying.

Egg oil is used in the production of shampoos as a nourishing additive to strengthen hair.

Egg oil, like vegetable oil, has a limited compatibility with grease liquids. No information on the compatibility of egg oil with powders was found in the literature/internet.

There is no need to **refine egg butter.**

3. Production of albumin.

Process 2.

After separation (extraction) of the egg oil from the egg powder (**Process 1**), pure egg white remains. It is dried under vacuum from residual solvent CH_2 Cl_2 and, after additional pasteurisation, packed in hermetically sealed containers.

Dry albumin is a stable product. It mixes under vigorous agitation with greasy, water-containing liquids.
Albumin is mixed with powders of organic and inorganic origin. **Albumin can be refined** after manufacture by converting it to water.

4. L-a-lecithin production.

The feedstock for this stage is the egg oil obtained in **Process 1**.

Egg oil (100kg) is fed into reactor R2 (V=0.6 m3) by pump from E1. Here also solvent S2 - acetone (400l) is fed from the container E2 under constant intensive stirring. A coloured wax-like precipitate falls out. After 1 hour the upper transparent light-yellow solution is pumped to the vacuum evaporator I2, where the egg oil devoid of lecithin is separated from the solvent S2. This egg oil (~85kg) from the lower part of the vacuum evaporator I2 is drained into a separate container E5, and the evaporated solvent S2 is condensed and collected in the intermediate container ES2. The solvent S2 is then pumped back to its original container E2 and reused in the process.

To the precipitate (waxy red-brown mass) remaining in the reactor R2 (washed twice in 50l with solvent S2 - acetone), solvent S3 - ethanol (50l) is fed from the container E3 under constant stirring. The obtained yellow transparent solution of L-a-lecithin is pumped to the vacuum evaporator I3, where L-a-lecithin is separated. The expelled solvent S3 is condensed and collected in the intermediate vessel ES3. The solvent S3 is then pumped to its original container E3 and reused in this process. L - a - lecithin from evaporator I3 after removal of residual solvent S3 is collected in container E6 and quickly poured at +40 C in an inert gas current (nitrogen, argon) into hermetically sealed containers.

Store in the fridge only!

L- a -lecithin can be converted into powder by repeated extraction of egg oil residues with acetone (total acetone consumption up to 15l per 1kg of pure product).

A more effective method is to use a mixture of two solvents - hexane/acetone at the rate of 50ml of hexane and 75ml of acetone per 100g of lecithin. In this case two phases are formed - the upper, lighter phase contains lecithin, and the lower, heavy phase contains phosphatides - egg oil residues.

It is possible to de - masculinise L- a -lecithin by extraction with supercritical CO .2

The purification of L-a -lecithin by ultrafiltration of a hexane solution through an ultrafilter with a pore width corresponding to molecules with a mol. mass of approximately 10,000 has been described in the literature. The solution passing through the filter is almost completely free of phosphatides.

L- a -lecithin is a good emulsifier, has stabilising and dispersing properties and can be mixed with grease liquids.

5. Lysozyme production.

The initial raw material for lysozyme extraction is the original native egg white. 1 litre of liquid egg white contains about 3 g of lysozyme.

One method described in the literature presents the direct crystallisation of lysozyme from the original liquid egg white in the presence of 5% NaCl.

It was found that in filtered and homogenised egg white at 4 C, pH=9.5 (adjusted with 1N KOH) and addition of 5% NaCl, lysozyme crystals are formed with 60-80% yield when standing for 3-4 days. Addition of a small amount of lysozyme crystals helps to initiate the process. The obtained crystalline product is dissolved in weak acetic acid (pH=4 -6) and all insoluble products are removed by centrifugation. The soluble product is recrystallised after adding 5% NaCl and adjusting the pH to 9.5 - 11.0. The recrystallisation can also be quickly completed by adding 5% sodium bicarbonate (pH=8.0-8.5) to the acidic solution.

Lyophilic drying completes the process.

The solubility of lysozyme in water is 10g/ml.

The product can be stored at -20 C without loss of activity for at least 4 years.

6. Production of multivitamins from eggs.

In reactor R2 after removal of lecithin alcohol solution a viscous red-brown mass remains - a mixture of multivitamins. It is suspended in 10 litres of hexane (petroleum ether with a boiling point of 40-60 C) and pumped or drained through the bottom hatch of the reactor into a vacuum evaporator. The solvent-free mixture of multivitamins is packed in hermetically sealed containers.

Printed by Books on Demand GmbH, Norderstedt / Germany